The People Problem:

Strengthening Cybersecurity Through Proper Training

By Dr. Jerrod Pickering

©2023

About the Author

Dr. Jerrod Pickering is an accomplished professional with a diverse background in education and cybersecurity. Holding a Doctorate in Instructional Technology from Texas Tech University, he has made significant contributions to the field through his expertise and experiences.

With a career spanning various roles in the education sector, Dr. Pickering has served as a high school teacher, high school principal, and school superintendent. These roles have provided him with valuable insights into the challenges faced by educational institutions in maintaining robust cybersecurity practices.

In his capacity as an external consultant, Dr. Pickering has traveled extensively throughout the United States, collaborating with financial institutions to enhance their security posture. His primary focus has been on identifying areas of non-compliance with FFIEC guidelines and other state and federal banking regulations. His deep understanding of the industry's unique security requirements has made him a trusted advisor in the financial sector.

Not limited to external consulting, Dr. Pickering has also worked as an internal IT auditor for one of the world's largest financial institutions. This experience allowed him to gain firsthand knowledge of the intricacies involved in safeguarding critical information within complex organizational structures.

To further augment his professional qualifications, Dr. Pickering is a Certified Information Systems Auditor licensed by ISACA, demonstrating his commitment to upholding industry standards and best practices. Additionally, he holds a Security+ certification from CompTIA, further validating his expertise in the field of cybersecurity.

Dr. Pickering's contributions extend beyond his practical work. He is an accomplished writer, with numerous articles authored by him covering a wide range of cybersecurity topics addressed in this book. His written works serve as valuable resources for professionals seeking practical insights and strategies to navigate the evolving cyber threat landscape.

In recognition of his expertise, Dr. Pickering has been invited to present at both regional and national conferences, as well as deliver insightful webinars. These platforms allow him to share his knowledge and educate audiences on the importance of cybersecurity awareness and the role of education in fostering a cyber-resilient workforce.

Combining his passion for education and cybersecurity, Dr. Pickering advocates for comprehensive workforce education to address the knowledge gaps that exist outside the IT space. His dedication to raising awareness and promoting best practices underscores his commitment to fostering a secure digital environment for individuals and organizations alike.

Through his extensive experience, qualifications, and unwavering dedication, Dr. Jerrod Pickering continues to make a profound impact on the field of cybersecurity, positioning himself as a leading authority in safeguarding critical information in an increasingly interconnected world.

Contents

Introduction

In a world where digital threats loom large and cyberattacks dominate headlines, it's easy to get caught up in the latest technological advancements and cutting-edge security tools. We pour vast amounts of money into fortifying our digital fortresses, convinced that we've safeguarded our organizations and ourselves against the perils of the digital age. But there's a problem—an insidious vulnerability that continues to persist despite our best efforts: the human factor.

Welcome to "The People Problem," a book that examines the pivotal role people play in cybersecurity. Through personal experiences, observations, and hard-hitting statistics, we'll delve into the uncharted territory where human fallibility meets the ever-evolving threat landscape. Prepare to embark on a journey that explores the profound impact of human behavior on our digital defenses.

Cybersecurity Ventures predicts that from 2021-2025, global cybersecurity spending will exceed $1.75 trillion while damage caused by cyber hacks will cost over $10 trillion annually, all while needing to protect 200 zettabytes of data. However, no matter how advanced our technology becomes, nor how impenetrable our firewalls or sophisticated our encryption methods, it is the people behind the screens who hold the keys to our collective security. Their actions, their decisions, and their awareness—or lack thereof—can make or break the resilience of our digital infrastructure.

But this is not a book intended solely for the cybersecurity professional or the tech-savvy enthusiast. While presenting some statistics, it is not a book filled with hard hitting numbers. Unless you have lived under a rock for the last 20 years, I think most people understand that cyber attacks are a problem. I was presenting on computer viruses and worms in high school back in 1996. This book is for everyone. It is for the CEO grappling with the complexities of securing their organization, the small business owner fighting to protect their livelihood, the parent navigating the perilous waters of online safety, and the individual seeking to safeguard their digital identity. This book is an invitation to explore the often overlooked but critical aspect of cybersecurity—ourselves.

Through personal anecdotes and stories drawn from the trenches of cyber warfare, I will shine a light on the human experience behind the headlines. But I won't stop there. Armed with tangible data and real-world case studies, together, we will confront the cold, hard truth and empower you with the knowledge and techniques needed to fortify the human link in the cybersecurity chain.

Approachable yet authoritative, "The People Problem" combines relatable narratives with a deep understanding of the cybersecurity landscape. My goal is to equip you, the reader, with the insights and tools necessary to tackle the challenges head-on and become an active participant in safeguarding our digital world.

So, whether you're a cybersecurity professional seeking a fresh perspective, an executive searching for strategies to protect your organization, or an individual concerned about your own digital safety, join me on this enlightening journey. Together, we'll navigate the intricate interplay between human behavior and the digital realm, understanding that by strengthening the people, we fortify our defenses against those who seek to exploit our vulnerabilities.

Chapter 1: The Human Factor in Cybersecurity

We often think of cybersecurity as a battle fought in the realm of code and technology. We picture hackers hunched over keyboards, launching sophisticated attacks against our digital fortresses. While these images hold some truth, there's a critical element that often goes overlooked—the human factor. It is within the minds, actions, and vulnerabilities of individuals that the true battleground of cybersecurity lies.

In our modern interconnected world, the impact of human behavior on cybersecurity cannot be underestimated. The *2023 Verizon Data Breach Investigation Report* indicates that 82% of all breaches involved the human element. Despite advancements in security tools and countless dollars spent on technological fortifications, we still find ourselves exposed to significant risks. The reason? People. It doesn't matter how much money we invest in the latest cyber defense systems. If our people aren't properly trained, they will always remain the weakest link in our security chain.

The implications of underestimating the human element in cybersecurity are staggering. Human error, negligence, and lack of awareness create avenues for cyber attackers to exploit. It's not uncommon to encounter scenarios where an unsuspecting employee unknowingly clicks on a malicious link, inadvertently providing hackers with access to sensitive data. Similarly, weak passwords, careless sharing of credentials, and the absence of two-factor authentication contribute to our vulnerabilities. These are the human elements that hackers exploit with precision.

But why are people often the weakest link? The answer lies in the unique characteristics of human nature. We are fallible, susceptible to manipulation, and prone to lapses in judgment. Cyber attackers recognize and capitalize on these inherent human traits. They employ tactics like social engineering, phishing, and psychological manipulation to bypass even the most robust technological defenses. It's a battle of wits, and the stakes are higher than ever before.

To fully comprehend the human factor, we need to explore real-life examples and anecdotes that highlight the significance of people-related issues in cyber attacks. Let's meet Sarah, a hardworking professional who fell victim to a cunning phishing attack. Sarah, like many of us, was busy juggling multiple tasks, her inbox brimming with emails demanding her attention. One seemingly innocuous email disguised as a routine message from her bank requested her login credentials. Oblivious to the dangers lurking in her inbox, she unwittingly handed over her personal information to cybercriminals.

Sarah's story is just one example of how a single click or a momentary lapse in judgment can have far-reaching consequences. The statistics surrounding human-related cybersecurity incidents are equally alarming. As previously mentioned, research shows that over 90% of successful cyber attacks can be attributed to human error or behavior. These numbers paint a vivid picture of the critical role people play in the cybersecurity landscape.

The good news is that with proper training and awareness, we can empower individuals to become our greatest defense against cyber threats. By investing in comprehensive and ongoing cybersecurity education, organizations and individuals can mitigate risks and bolster their resilience. Training programs should go beyond the basics of password management and delve into topics like recognizing phishing attempts, identifying suspicious links, and understanding the importance of data privacy.

In the upcoming chapters, we will explore various attack vectors, such as phishing, ransomware, and social engineering. Each chapter will delve into personal stories, backed by real-world statistics and case studies, to emphasize the impact of these attacks. Equally important, we will provide you with practical and effective training techniques to strengthen your defenses. By arming ourselves with knowledge and fostering a culture of cybersecurity awareness, we can shift the paradigm and turn the human factor from a vulnerability into a formidable line of defense.

In the following chapters, we'll dive deeper into specific attack vectors, uncovering their inner workings and sharing real-life stories that will shed light on their consequences.

Chapter 2: Phishing Attacks

Imagine

Imagine receiving an email that appears to be from your bank, urging you to update your account information urgently. The email seems legitimate, complete with the bank's logo and professional language. What would you do? This scenario is an example of a phishing attack—a clever and deceitful tactic employed by cybercriminals to trick unsuspecting individuals into revealing sensitive information. In this chapter, we'll explore the world of phishing attacks, sharing personal stories, revealing alarming statistics, and equipping you with effective training techniques to combat this pervasive threat.

Phishing attacks continue to be one of the most prevalent and successful forms of cyber attacks. *Cybersecurity Ventures* reports in their 2020 almanac that phishing attacks accounted for about 90% of all data breaches worldwide. The techniques used by cybercriminals have evolved, becoming increasingly sophisticated and difficult to detect. They exploit our trust in familiar organizations, leveraging the power of social engineering to manipulate our emotions and entice us to take actions that compromise our security.

Scenario

Take the case of Mark, a diligent employee who received an urgent email from his company's HR department. The email stated that due to a recent security breach, all employees were required to reset their passwords immediately by clicking on a provided link. Concerned about the security of his account, Mark complied without hesitation, unknowingly providing cybercriminals with his login credentials. This seemingly innocent email turned out to be a well-crafted phishing attempt, leaving Mark's personal and company data vulnerable.

Unfortunately, Mark's story is not an isolated incident. Statistics reveal the alarming success rate of phishing attacks. According to industry reports, approximately 90% of successful cyber attacks begin with a phishing email. In addition, over 30% of phishing emails are opened by their recipients, and about 12% of those recipients click on the malicious links or download harmful attachments. 12% may not sound like much initially. However, when an attacker sends out 10,000 emails that is 1,200 people that click on the link. If each one of those clicks results in a loss of $500, then the attacker will get $600,000 off one mass email. It puts that 12% into a completely different perspective. These numbers underscore the pressing need for robust training and awareness to counter this pervasive threat.

Personal Experience Regarding Phishing

During my tenure as an external consultant, I had the privilege of conducting phishing attack tests on behalf of my clients. It took place in 2016, a time when people were becoming increasingly aware of common phishing techniques such as misspellings and suspicious email addresses. To challenge the status quo and raise the bar, our team embarked on a mission to craft sophisticated phishing emails that would truly test the vigilance of recipients.

We aimed to create a sense of familiarity and trust by incorporating personal names that recipients might recognize. However, the underlying email addresses were carefully designed as dummy accounts, ensuring no real harm would come to the recipients. The core objective was to educate rather than exploit. Upon clicking any of the embedded links within the email, individuals were redirected to a landing page where we revealed that they had fallen victim to a phishing attempt.

To enhance the learning experience, we prepared a concise one to two-minute video that served as a quick-hit educational resource on how to identify phishing emails. This video emphasized essential techniques and provided practical insights to help recipients distinguish between genuine communications and malicious attempts. Our intention was never to shame or embarrass but rather to empower individuals with the knowledge and skills necessary to safeguard themselves and their organizations against future threats.

This endeavor served as a valuable exercise in highlighting the significance of continuous education and awareness in the ever-evolving realm of cybersecurity. It underscored the importance of staying ahead of malicious actors by equipping individuals with the tools to identify and mitigate phishing attacks. By blending practical techniques with an understanding of human psychology, we aimed to foster a culture of cyber resilience and empower individuals to become the first line of defense against these pervasive threats.

Phishing Training

To protect ourselves and our organizations from phishing attacks, it is crucial to cultivate a vigilant and informed mindset. Here are some effective training techniques to strengthen your defense against phishing:

- Awareness and Education: Regularly educate employees, family members, and yourself about the latest phishing techniques, warning signs, and best practices for safe online behavior. Teach them to be skeptical of unsolicited emails, especially those requesting sensitive information.

- Suspicion and Verification: Encourage a healthy level of skepticism. Advise individuals to verify the legitimacy of emails or requests by contacting the organization directly through trusted channels. Emphasize the importance of never clicking on suspicious links or downloading attachments from unknown sources.

- Strong Passwords and Two-Factor Authentication: Promote the use of strong, unique passwords for all online accounts and emphasize the importance of not sharing passwords. Encourage the adoption of two-factor authentication, which provides an extra layer of security.

- Phishing Simulations: Conduct regular phishing simulation exercises within organizations and households to test individuals' ability to identify and respond to phishing attempts. These simulations offer valuable training opportunities and help reinforce awareness.

- Reporting Mechanisms: Establish clear reporting mechanisms for suspected phishing attempts. Encourage individuals to report suspicious emails or incidents promptly, fostering a collaborative and proactive approach to cybersecurity.

Remember, combating phishing attacks requires a collective effort. By implementing these training techniques and fostering a culture of cybersecurity awareness, we can significantly reduce the success rate of phishing attempts and protect ourselves and our organizations from the devastating consequences of falling victim to these deceptive tactics.

In the next chapter, we'll explore another insidious cyber threat: ransomware attacks. We'll share compelling stories, present eye-opening statistics, and equip you with effective strategies to mitigate the risks associated with this increasingly prevalent form of attack.

Chapter 3: Ransomware Attacks

Imagine

Picture waking up one morning to find that all the critical files on your computer or within your organization are encrypted and inaccessible. A chilling message appears on your screen, demanding a hefty ransom in exchange for the decryption key. This nightmare scenario represents the devastating reality of ransomware attacks—the insidious cyber threat that has wreaked havoc on individuals and organizations alike. In this chapter, we'll delve into the world of ransomware attacks, sharing compelling stories, revealing eye-opening statistics, and equipping you with effective strategies to mitigate the risks and protect yourself from this escalating danger.

Ransomware attacks have become alarmingly prevalent in recent years, causing significant financial losses and operational disruptions. The *2023 Verizon Data Breach Investigations Report* hypothesizes that the reason for this rise is because it is inexpensive to acquire and since its primary attack vector is email, it relatively inexpensive to deliver compared to other types of cyber attacks. Cybercriminals behind these attacks exploit vulnerabilities in our systems and human behavior, leveraging advanced encryption techniques to hold our valuable data hostage. Their primary motive? Financial gain.

Let's meet Laura, a small business owner who fell victim to a ransomware attack. One seemingly harmless click on a malicious email attachment led to the complete encryption of her company's sensitive data. The attackers demanded a substantial ransom in exchange for the decryption key, leaving Laura with a difficult choice: pay the ransom and hope for the return of her data or face the potentially catastrophic consequences of losing it forever. Laura's story highlights the devastating impact of ransomware attacks, where individuals and organizations find themselves at the mercy of cybercriminals.

The scope and magnitude of ransomware attacks are truly alarming. According to recent reports, the number of ransomware attacks has increased by over 150% in the past two years, making it one of the fastest-growing cyber threats. Furthermore, the costs associated with these attacks, including ransom payments, remediation, and reputational damage, are projected to reach billions of dollars annually.

Personal Story Regarding Ransomware

Early on in my career as a dedicated cybersecurity consultant, I encountered a distressing incident that vividly demonstrated the real-world consequences of falling prey to cybercriminals. It all started with a phone call from a concerned relative who uttered those dreaded words, "Jerrod, those persistent Russians have struck again."

Puzzled, I inquired further, seeking to understand the nature of the predicament. She proceeded to explain that her assistant, in a moment of unsuspecting innocence, had clicked on a seemingly harmless link disguised as an Amazon tracking number. Little did they know, this single click had triggered a malicious ransomware attack that swiftly engulfed their office.

With their screens abruptly rendered blank, a chilling message materialized, demanding a ransom of half a Bitcoin in exchange for the safe return of their files. Faced with this dire situation, my relative turned to me, seeking guidance on how to navigate the treacherous waters they found themselves in.

Understanding the gravity of the situation, I immediately advised her to disconnect her machine from the network, a crucial step in preventing further damage. I then proceeded to explain the intricacies of ransomware and the tough road that lay ahead. As I painted a realistic picture of the weeks to come, a sense of disappointment washed over her face, realizing the long and arduous journey that awaited her and her office.

In the ensuing weeks, their office had to grapple with the immense challenges brought forth by the ransomware attack. Payroll had to be painstakingly processed by hand, reverting to manual systems of the past. Acquiring Bitcoin and navigating the complexities of digital wallets became a necessity. They were compelled to enlist the expertise of an external agency to thoroughly cleanse and restore their compromised computers. Even with backups in place, the process of recovering data proved to be an uphill battle, exacting a hefty toll on their resources and resilience.

This poignant tale serves as a reminder of the critical importance of cybersecurity preparedness. It highlights the pressing need for organizations and individuals alike to fortify their defenses, cultivating a proactive mindset that wards off potential threats. Through proper training, robust security measures, and a culture of vigilance, we can shield ourselves from the devastating impact of ransomware attacks, preserving our precious data and safeguarding our livelihoods.

Ransomware Training

To mitigate the risks posed by ransomware attacks, it is crucial to adopt a proactive and multi-layered defense strategy. Here are some effective strategies to protect yourself and your organization:

- Regular Data Backup: Maintain secure and up-to-date backups of your critical data on separate storage systems or in the cloud. Regularly test the integrity and accessibility of these backups to ensure their effectiveness in the event of an attack.

- Robust Security Measures: Implement comprehensive security measures, including strong firewalls, up-to-date antivirus software, and intrusion detection systems. Keep all software and operating systems patched and updated to address known vulnerabilities.

- Employee Awareness and Training: Educate employees about the risks of ransomware attacks and provide regular training on safe online practices. Teach them to be cautious of suspicious emails, avoid clicking on unverified links or downloading attachments from unknown sources, and report any suspicious activities promptly.

- Network Segmentation: Segment your network to isolate critical systems and data, reducing the impact of a potential ransomware attack. Limit access privileges to only those individuals who require them for their roles.

- Incident Response Plan: Develop a comprehensive incident response plan that outlines clear procedures for responding to a ransomware attack. This plan should include steps for isolating infected systems, notifying appropriate authorities, and recovering data from backups.

- Regular Testing and Simulation: Conduct regular testing and simulation exercises to evaluate the effectiveness of your security measures and incident response plan. Identify any weaknesses or gaps in your defenses and address them promptly.

By adopting these strategies and maintaining a proactive stance, you can significantly reduce the risks associated with ransomware attacks. Remember, prevention and preparedness are key when it comes to combating this evolving threat landscape.

In the next chapter, we'll explore another dangerous attack vector: SMishing attacks. We'll uncover the intricacies of these text message-based attacks and share real-life stories from the field.

Chapter 4: Smishing/Vishing Attacks

Imagine

Imagine receiving a text message from what appears to be your bank, alerting you to suspicious activity on your account. The message urges you to call a provided phone number immediately to resolve the issue. Unbeknownst to you, this text message is a carefully crafted SMishing attack, aiming to deceive you into providing your personal information to fraudsters.

In today's digital landscape, cyber attackers are relentless in their pursuit of exploiting our vulnerabilities. They constantly adapt their tactics to exploit the ways in which we communicate and interact with technology. In this chapter, we'll dive into the world of SMishing and Vishing attacks—clever schemes that leverage text messages (SMS) and voice calls (Vishing) to deceive individuals and steal sensitive information. Through personal stories, eye-opening statistics, and effective strategies, I'll equip you with the knowledge to recognize and protect yourself from these insidious forms of attack.

SMishing and Vishing attacks share a common objective: to trick individuals into revealing confidential information, such as account credentials, credit card details, or personal data. These attacks rely on social engineering techniques, preying on our trust and exploiting our natural inclination to respond to messages and phone calls.

Scenario

Let's meet David, a victim of a Vishing attack. One day, he received a call from someone claiming to be from a reputable tech support company. They informed him that his computer had been compromised and convinced him to provide remote access to fix the issue. Unbeknownst to David, he had fallen victim to a Vishing attack, inadvertently granting the attackers unrestricted access to his personal information and sensitive files.

Personal Vishing Story

In my early days as a cybersecurity consultant, I had a particularly intriguing experience that shed light on the importance of employee awareness. One of the services we offered to banks involved our team attempting to acquire passwords from their employees for testing purposes. We employed various methods, including internal vulnerability scans and penetration tests, but one of the most fascinating techniques was the creation of a fake website.

To execute this approach, we meticulously scraped images and other elements from a bank's legitimate website and then reserved a domain to construct a deceptive employee portal. Our goal was to trick individuals into entering their credentials on this bogus platform. To enhance the authenticity of our ruse, we even went as far as disguising our caller ID to make it appear as though we were calling from an internal number at the bank.

During one memorable incident, I called a bank located in a Texas lakeside area. Armed with knowledge about the region, I engaged a gentleman in conversation by introducing myself as a new IT tech support member working on the implementation of an internal employee portal. Aware that security-conscious employees would verify the legitimacy of my claim, I commended his cautious approach as he mentioned wanting to confirm my identity information with his boss.

Taking advantage of the brief waiting period, I initiated a friendly conversation and inquired about his recent fishing endeavors. Our conversation flowed as he provided enthusiastic updates on his fishing experiences. Eventually, he expressed frustration that his boss had not yet responded to his inquiry. Seizing the opportunity, I suggested he visit the website to take a look while we continued our conversation.

Our meticulously crafted website resembled the bank's genuine portal, featuring a username and password box prominently displayed in the center. After a few minutes, with his boss still unresponsive, I casually directed his attention to the username and password box, assuring him that I did not require his actual credentials. Instead, I asked him to enter the same username and password he used to log in to his computer.

Curiosity piqued, he cautiously proceeded, noting the authenticity of the website. He entered his username and password, unaware that our configuration led to a cleverly disguised "404 error" page, while we discreetly captured his credentials in the background. Unfazed by the initial error, he attempted to reenter his information, falling into our trap once again.

With a touch of regret, I informed him that we clearly had a lot of work ahead, suggesting that it would be a long weekend for me. Expressing gratitude for his time, I bid him farewell and concluded the call. Immediately afterward, I contacted our bank contact to report the compromised username and password, stressing the significance of this incident as a valuable learning opportunity for the organization. Unfortunately, I never learned what transpired afterward, hoping that the bank effectively leveraged this experience to strengthen its security measures through the training of its employees.

This captivating story highlights the significance of employee awareness and the critical role training plays in combating cyber threats. It underscores the vulnerability of even the most cautious individuals and emphasizes the need for organizations to invest in comprehensive cybersecurity education to fortify their defenses against malicious actors.

The prevalence of SMishing and Vishing attacks is alarming but often under reported. The *2023 Verizon DBIR* suggests that the reason for this, is often times a phishing link is sent by text to a mobile device and when it is clicked, the statistics is reported as a phishing attempt. The data provided indicates that at least 58% of mobile devices clicked a malicious URL at least once. This number underscores the urgent need for awareness and proactive measures to safeguard ourselves against these evolving threats.

Smishing/Vishing Training

To mitigate the risks associated with SMishing and Vishing attacks, it's crucial to adopt a comprehensive defense approach. Here are some effective strategies to protect yourself:

- Verify the Source: Be cautious when receiving unexpected text messages or phone calls. Never provide personal information or financial details without independently verifying the legitimacy of the request. Contact the organization directly using official contact information to confirm the authenticity of the message or call.

- Be Skeptical: Maintain a healthy level of skepticism when encountering urgent or alarming messages or calls. Fraudsters often create a sense of urgency or exploit emotions to prompt immediate action. Remember, legitimate organizations will not pressure you to disclose sensitive information abruptly.

- Never Click on Suspicious Links: Avoid clicking on links provided in text messages or emails from unknown sources. These links may direct you to malicious websites or prompt the download of malware onto your device.

- Protect Personal Information: Be cautious about sharing personal information, such as social security numbers, passwords, or financial details, over the phone or via text messages. Legitimate organizations typically won't request sensitive information through these channels.

- Educate and Train: Provide education and training on SMishing and Vishing attacks to employees, family members, and yourself. Teach individuals to recognize common tactics used in these attacks and empower them to report any suspicious messages or calls promptly.

- Implement Call Filtering and Anti-Spam Solutions: Utilize call filtering services and anti-spam solutions that can help identify and block potential Vishing calls or fraudulent text messages.

By adopting these strategies and maintaining a vigilant mindset, you can significantly reduce the risks associated with SMishing and Vishing attacks.

However, prevention is just one part of the equation. It's also essential to have a response plan in place in case you do encounter such attacks. Here are some additional measures to consider:

- Incident Response Plan: Develop an incident response plan that outlines the steps to take in the event of a SMishing or Vishing attack. This plan should include instructions on how to report the incident, how to limit the potential damage, and whom to contact for assistance.

- Stay Informed: Keep yourself updated on the latest SMishing and Vishing attack techniques. Stay informed about the new tactics cybercriminals employ to deceive individuals. Regularly follow reputable cybersecurity news sources and subscribe to threat intelligence feeds to stay ahead of emerging threats.

- Multi-Factor Authentication (MFA): Enable multi-factor authentication wherever possible, especially for sensitive accounts such as email, banking, or social media. MFA adds an extra layer of security by requiring an additional verification step, such as a unique code sent to your phone, in addition to your password.

- Anti-Malware and Anti-Phishing Solutions: Install and regularly update reputable anti-malware and anti-phishing software on your devices. These tools can detect and block malicious content, helping to prevent SMishing and Vishing attacks.

- Trust Your Instincts: Trust your gut instinct when something feels off. If a message or call seems suspicious, do not engage further. Disconnect the call, delete the text message, or simply ignore it. It's better to be safe than sorry.

Remember, combating SMishing and Vishing attacks requires a combination of awareness, education, and proactive measures. By implementing these strategies and maintaining a cautious approach, you can significantly reduce the likelihood of falling victim to these deceptive tactics.

In the next chapter, we'll explore another critical aspect of cybersecurity—social engineering. We'll uncover the psychological manipulation techniques employed by cyber attackers and provide you with practical insights to recognize and protect yourself from these manipulative strategies.

Stay informed, stay cautious, and together, let's build a stronger defense against evolving cyber threats.

Chapter 5: In-Person Social Engineering

Imagine

Imagine a scenario where an individual dressed as a repair technician gains access to an office building by convincing an unsuspecting employee that they need to perform urgent maintenance. Once inside, the attacker may tamper with systems, plant malware-infected devices, or gather sensitive information. This is just one example of how in-person social engineering can bypass even the most robust technological defenses, targeting the human element—the weakest link in cybersecurity.

When we think about cyber attacks, we often imagine a faceless hacker sitting behind a computer screen. It is true that most forms of social engineering is weaved through the various attack vectors it is important to not underestimate the power of an in-person social engineering attempt. Some of the most effective and dangerous attacks occur through in-person interactions, leveraging the power of social engineering. In this chapter, we'll delve into the realm of in-person social engineering, where attackers manipulate human psychology and exploit trust to gain unauthorized access to sensitive information. Through personal stories, eye-opening statistics, and effective strategies, we'll shed light on these deceptive tactics and empower you to protect yourself from this deceptive form of attack.

In-person social engineering relies on the art of deception and manipulation, preying on our inherent trust in others. Attackers may impersonate legitimate individuals or pose as authority figures, exploiting our desire to be helpful or compliant. They use a variety of tactics to manipulate human behavior, such as exploiting vulnerabilities, appealing to emotions, or creating a sense of urgency. While in-person attempts places a higher risk on the attacker, it still is an attack vector that plays with a person's emotions to elicit a response.

Scenario

Meet Alex, a victim of an in-person social engineering attack. While working at a coffee shop, he struck up a conversation with a friendly stranger who happened to be sitting nearby. Over time, they built a rapport, and Alex gradually disclosed personal details about his work and daily routine. Unbeknownst to him, this stranger was an attacker skilled in eliciting information. Armed with the knowledge gained through their conversations, the attacker later impersonated a coworker of Alex's, gaining unauthorized access to his workplace and sensitive information. Alex's story serves as an important reminder of the risks we face in seemingly harmless interactions.

The prevalence and impact of in-person social engineering attacks are concerning. Statistics from the *2023 Verizon DBIR* suggest that approximately 60% of successful data breaches involve social engineering tactics, with in-person attacks being a significant component. Moreover, research indicates that human error contributes to nearly 95% of security incidents. These statistics underscore the urgent need for awareness and proactive measures to protect against in-person social engineering attacks.

Personal Story Regarding In Person Attempts

During my tenure as a cybersecurity consultant, I participated a captivating story that shed light on the vulnerabilities of social engineering within a bank's security protocols. One of the key tasks assigned to our team was to perform social engineering tests on bank branches, simulating scenarios where we posed as telecom workers requiring access to the server room. This particular incident took place on a scorching July afternoon, following a series of brownouts throughout the day.

Arriving at the bank just 30 minutes before closing time, I approached the first teller I encountered and apologized for the delay, mentioning that I had to drive from another city to reach their branch. Sensing the frustration caused by the sluggish internet service that day, I made a casual remark about the bank's Internet issues, a complaint commonly heard among many individuals. The teller concurred, acknowledging the slow connection they had experienced. Seizing the opportunity, I requested if she could guide me to the equipment room, I would work on resolving their connectivity problems.

Following protocol initially, the teller asked for my driver's license and called her manager over to review the identification as well. While the manager scrutinized my license, I engaged them in friendly conversation, empathizing with the challenges of dealing with brownouts at the end of a long day. If the manager continued to follow protocol, they should have contacted the main branch to verify that I was supposed to be there. This manager did not complete this second step, but did initially accompanied me to the switch room, visibly eager to wrap up their day's responsibilities. Sensing their urgency, I insisted they could leave me unattended and assured them I would find my way out. Grateful for the offer, the manager left me alone in the switch room.

Inside the room, I treaded carefully, keenly aware of the opportunities for mischief that lay before me. As a cybersecurity professional, I understood the potential risks of unauthorized access. However, in this test scenario, my goal was to assess the bank's security practices rather than exploit them. Hence, I refrained from any intrusive actions such as inserting USB drives or attempting network infiltration. Instead, I merely familiarized myself with the surroundings, observing the setup and taking mental notes.

Upon leaving the switch room, I returned to my car to document the details of the encounter. However, as I typed out my notes, a realization struck me—I had neglected to capture photographic evidence of my unescorted presence in the switch room, a crucial piece of documentation to support our findings. Determined to rectify this oversight, I reentered the branch and approached the employee once again. Explaining that

I had left a piece of equipment behind in the switch room, I requested permission to retrieve it.

To my surprise, the employee readily agreed and allowed me to return to the switch room unaccompanied. With my camera in hand, I discreetly captured the necessary photographs to substantiate our report. Satisfied with the evidence gathered, I promptly reported the findings to our bank contact at the main location. Regrettably, as is often the case with these tests, I never learned the specific outcomes or measures taken by the bank in response to this incident. Nonetheless, I remain hopeful that this experience served as an educational opportunity to reinforce their security practices.

This story underscores the critical role of social engineering awareness within the banking sector. It emphasizes the importance of employee vigilance and the need for robust protocols to verify the identities and intentions of individuals seeking access to sensitive areas. By sharing such experiences, we aim to shed light on the intricacies of social engineering and inspire organizations to fortify their defenses against this ever-evolving threat landscape.

Social Engineering Education Suggestions

To defend against in-person social engineering attacks, it's crucial to cultivate a security-minded culture and employ effective strategies. Here are some key steps to consider:

- Awareness and Education: Educate yourself and your organization about the techniques and tactics employed in in-person social engineering attacks. Train employees to recognize common manipulation techniques and emphasize the importance of verifying identities and following proper protocols.

- Vigilance and Trust Your Instincts: Develop a healthy sense of skepticism and trust your instincts when interacting with unfamiliar individuals or facing unexpected requests. If something feels off or raises suspicion, take a step back and question the situation.

- Verify Identities and Authorization: Always verify the identities and authorization of individuals before granting access to restricted areas or sensitive information. Ask for proper identification or contact a designated authority to confirm their legitimacy.

- Implement Access Controls: Utilize access control systems, such as keycards or biometric authentication, to limit unauthorized access to physical spaces. Regularly review and update access permissions to ensure they align with current requirements.

- Confidentiality and Data Protection: Treat sensitive information with the utmost confidentiality, both in verbal and written communications. Be mindful of discussing sensitive matters in public spaces and adopt proper data protection measures, such as secure document disposal.

- Incident Reporting and Response: Establish clear procedures for reporting and responding to suspected or confirmed incidents of in-person social engineering. Encourage employees to report any suspicious interactions or requests promptly. Designate a responsible team or individual to handle incident response, conduct investigations, and implement appropriate remediation measures.

- Social Engineering Testing: Conduct regular social engineering testing within your organization to identify potential vulnerabilities and educate employees on common tactics. This can involve simulated scenarios where employees are tested on their ability to recognize and resist social engineering attempts.

- Physical Security Measures: Implement robust physical security measures, such as surveillance cameras, access logs, and visitor registration systems. These measures can deter unauthorized individuals and provide valuable evidence in case of an incident.

- Ongoing Training and Awareness: Provide ongoing training and awareness programs to reinforce security best practices and keep employees informed about the latest social engineering techniques. This can include interactive workshops, educational materials, and simulated exercises to enhance resilience against in-person social engineering attacks.

- Trustworthy Relationships and Verification: Foster trusted relationships with external vendors, contractors, and service providers. Maintain clear communication channels to verify requests or changes in procedures before granting access or sharing sensitive information.

By adopting these strategies and instilling a security-conscious mindset, you can strengthen your defense against in-person social engineering attacks. Remember, vigilance and skepticism are key, but it's equally important to cultivate an environment where employees feel comfortable reporting suspicious activities without fear of retribution.

In the next chapter, we will explore the vital role of penetration testing, also known as ethical hacking, in identifying vulnerabilities and fortifying your organization's cybersecurity defenses. We'll uncover the process of simulating real-world attacks to uncover weaknesses and provide you with insights into leveraging penetration testing effectively.

Stay alert, stay informed, and together, let's outsmart the social engineers.

Chapter 6: Penetration Testing and Vulnerability Assessments

Imagine

Imagine a scenario where a company believes its network and applications are secure, only to discover that sensitive customer data has been compromised by a cybercriminal. Such incidents can result in significant financial losses, damage to reputation, and legal consequences. However, with a well-executed penetration test, vulnerabilities can be exposed before they are exploited, allowing organizations to implement the necessary safeguards.

In a world where cyber threats continue to evolve rapidly, organizations must stay one step ahead of malicious actors. To truly understand the vulnerabilities in their systems and fortify their defenses, proactive measures are essential. Enter penetration testing, a vital practice that mimics real-world attacks to identify weaknesses and secure digital infrastructure. In this chapter, we will explore the world of penetration testing and ethical hacking, delving into its importance, methodologies, and best practices. Through personal stories, eye-opening statistics, and actionable strategies, I will empower you to leverage penetration testing effectively and protect your digital assets.

Penetration testing, often referred to as ethical hacking, involves authorized and controlled attempts to breach the security of an organization's systems. Skilled professionals, known as ethical hackers or penetration testers, employ various techniques to simulate real-world attacks, exposing vulnerabilities and providing actionable recommendations to mitigate risks.

Scenario

Meet David, an experienced penetration tester. He was hired by a financial institution to assess their security posture and identify potential weaknesses. Through a systematic and rigorous evaluation, David uncovered critical vulnerabilities that could have led to unauthorized access, data breaches, and financial losses. Thanks to his findings, the organization was able to patch the vulnerabilities, enhance their security measures, and prevent potential disasters.

The importance of penetration testing cannot be overstated. According to a 2022 article in "Forbes," organizations are increasingly understanding the need for penetration testing to help secure their network. Furthermore, penetration testing can significantly reduce the average cost of a data breach, saving organizations millions of dollars in potential damages.

Personal Vulnerability Assessment Story

As an external consultant specializing in penetration testing, our team was entrusted with the critical task of conducting the initial step in pen testing known as reconnaissance for various banks. This initial phase involved scanning the banks' external IP addresses to uncover potential vulnerabilities. Additionally, we were contracted to perform an internal vulnerability assessment (IVA), in which we placed a pc inside of the bank's network and scanned for internal vulnerabilities. I vividly recall a noteworthy incident that occurred during one such engagement when we stumbled upon a concerning vulnerability within the bank's printer infrastructure.

At the time, certain printers in the bank were exposed to a relatively new vulnerability known as a POODLE attack. Our team meticulously documented this finding and presented it to the bank's IT management as part of our comprehensive report. However, to our surprise, management seemed skeptical about the level of risk associated with this particular vulnerability. They challenged us to demonstrate how it could be exploited and requested tangible evidence to justify their concerns.

Determined to illustrate the potential consequences of overlooking this vulnerability, one of our highly skilled team members took charge. With expert precision, they proceeded to showcase how the POODLE attack could compromise the encryption between PCs and the printers. By decrypting supposedly secure data, confidential information became susceptible to interception and exploitation. This eye-opening demonstration left no room for doubt, prompting IT personnel to swiftly acknowledge the gravity of the situation.

The immediate remediation efforts that followed were a testament to the power of practical examples and the pivotal role humans play in the security landscape. While this specific threat originated from our IVA, it serves as a stark reminder that even seemingly minor configuration settings—handled by individuals—can have far-reaching implications for the overall security posture of an organization. By shedding light on the interconnectedness between technology and human behavior, we aim to empower individuals and organizations to better safeguard their critical assets.

PenTest Training

To leverage penetration testing effectively, organizations should consider the following best practices:

- Define Clear Objectives: Clearly define the goals and objectives of the penetration test. Identify specific systems, applications, or areas of concern that need to be assessed. This ensures that the testing aligns with organizational priorities and addresses potential risks.

- Engage Experienced Professionals: Work with experienced and certified penetration testers who possess the necessary expertise and qualifications. This ensures that the tests are conducted effectively, following established methodologies and industry best practices.

- Comprehensive Scope: Determine the scope of the penetration test, encompassing various aspects of the organization's digital infrastructure. This includes networks, applications, databases, wireless networks, and physical locations, among others. A comprehensive scope provides a holistic view of vulnerabilities and potential entry points for attackers.

- Methodical Testing Approach: Testing should follow a structured and methodical approach. This typically includes reconnaissance, vulnerability scanning, exploitation, and post-exploitation analysis. By following a systematic process, ethical hackers can identify vulnerabilities and simulate real-world attack scenarios.

- Collaboration and Communication: Foster collaboration between the organization's internal IT and security teams and the external penetration testing team. Maintain open lines of communication to ensure a thorough understanding of the organization's systems and to facilitate timely sharing of findings and recommendations.

- Documentation and Reporting: Document all the steps taken during the penetration testing process, including methodologies, tools used, vulnerabilities discovered, and their potential impact. Prepare a comprehensive report that clearly outlines the findings, risks, and actionable recommendations for remediation.

- Remediation and Follow-Up: Act promptly on the recommendations provided in the penetration testing report to address the identified vulnerabilities. Prioritize remediation efforts based on the severity and potential impact of each vulnerability. Regularly review and track progress to ensure that recommended fixes are implemented effectively.

- Continuous Testing and Improvement: Penetration testing should be an ongoing process rather than a one-time event. Regularly schedule and conduct follow-up tests to validate the effectiveness of implemented security measures and identify any new vulnerabilities that may have emerged. Stay up to date with evolving threats and adapt your testing approach accordingly.

- Compliance and Regulatory Requirements: Consider industry-specific compliance regulations and standards when conducting penetration testing. Ensure that your testing practices align with the required guidelines and fulfill any legal obligations.

- Learn from the Findings: Treat penetration testing as a valuable learning opportunity. Use the insights gained from the testing process to enhance your organization's overall security posture. Share key takeaways with relevant stakeholders to raise awareness and promote a culture of cybersecurity within your organization.

By embracing penetration testing and ethical hacking as proactive security measures, organizations can significantly reduce their risk of falling victim to cyber attacks. However, it's important to remember that these types of tests should always be conducted within legal and ethical boundaries, with the proper authorization and consent from relevant stakeholders.

In the next chapter, we will explore another unconventional attack vector: dumpster diving. We will uncover the surprising risks associated with discarded information and explore strategies to mitigate these threats effectively.

Stay vigilant, stay secure, and together, let's keep our digital assets safe.

Chapter 7: Dumpster Diving and Information Leakage

Imagine

Imagine a scenario where an organization disposes of sensitive documents without proper consideration for their content. A malicious individual rummages through their trash and discovers financial statements, customer records, or employee credentials. Armed with this valuable information, the attacker can wreak havoc, leading to identity theft, financial fraud, or even corporate espionage. Dumpster diving exposes the stark reality that improper disposal of information can have dire consequences.

When it comes to cybersecurity, we often focus on sophisticated cyber attacks and technological vulnerabilities. However, some of the most alarming breaches occur through a surprisingly low-tech method: dumpster diving. In this chapter, we will explore the risks associated with dumpster diving and information leakage, shedding light on the potential consequences of discarded information falling into the wrong hands. Through personal stories, eye-opening statistics, and actionable strategies, I will empower you to protect your organization's sensitive data and mitigate the risks associated with dumpster diving.

Scenario

Meet Sarah, an employee who inadvertently contributed to an information leakage incident. Rushed to meet a deadline, she discarded printed documents containing sensitive client information in a regular trash bin. Unbeknownst to her, a dumpster diver seized the opportunity and retrieved those discarded papers. The aftermath of this incident resulted in financial losses for both the organization and the affected clients. Sarah's story highlights the importance of proper disposal practices and the need for awareness of information leakage risks.

The prevalence and impact of dumpster diving incidents are alarming. Although most information acquisition is through electronic media, the *2023 Verizon DBIR* notes that approximately 30% of data breaches involve some form of physical document compromise. Surprisingly, many of these breaches could have been prevented with proper information disposal practices. This underscores the urgent need for organizations to prioritize the secure handling and disposal of sensitive data.

Personal Story Regarding Information Leakage

During my time as an external consultant, I had the opportunity to conduct various assessments that shed light on information leakage and the crucial role humans play in safeguarding sensitive data. Allow me to share a couple of compelling stories that emphasize the importance of protecting confidential information.

In one instance, our team was engaged by a medium-sized bank with branches spread across different cities. Recognizing that auditors are often seen as adversaries, we focused on building rapport with the bank staff to foster a collaborative environment. Through casual conversations over several days, our main contact felt comfortable enough to inquire about our experience with testing clean desk policies.

A Clean Desk Policy is a set of guidelines that encourage employees to maintain a tidy workspace by securely storing and removing sensitive documents and materials when not in use. It helps protect confidential information, prevent unauthorized access, and foster a culture of security awareness within the organization.

Eager to assist, we agreed to conduct a clean desk inspection after hours since our contact had noticed multiple compliance issues but lacked the authority to enforce the policy. Arriving later in the evening after the cleaning staff had left, we embarked on the assessment. What we discovered left us astounded.

On the desks, we found open files containing people's confidential information, such as Social Security numbers, tax records, and bank account details. Additionally, we stumbled upon signed cashier's checks left exposed for anyone to grab. To our disbelief, we even uncovered the master ATM card, which controls all ATM functions, along with its PIN number conveniently written on the card's sleeve. It was a glaring example of how information leakage could occur due to lax security practices.

In another scenario, we visited a small-town bank that, despite its size, was subject to the same federal and state regulations as larger institutions. These banks typically place great emphasis on community involvement and may overlook certain security measures that could hinder their community-oriented approach.

Before each engagement, I would browse the bank's social media accounts to gauge the bank's climate and culture. During one such review, I stumbled upon a picture that initially seemed harmless—a group of employees in the bank's lobby, congratulating Mr. and Mrs. Smith on their new loan. However, upon closer inspection, I noticed that in the background, a teller's PC screen displayed a customer's confidential information, visible to anyone who viewed the picture. It highlighted an unexpected avenue through which information leakage could occur.

As I continued scrolling through the feed, I encountered another concern. Small-town banks often foster close relationships with local schools, sponsoring billboards or scoreboards for sports teams and participating in school activities. On this particular bank's social media page, I noticed pictures from a field trip where kindergarten and first-grade students were guided through the bank, including behind the teller line.

While I had no issue with the children exploring behind the teller line, it struck me that adults—teachers and parent volunteers—also accompanied the students. This raised concerns about potential access to confidential information during regular working hours. To address this, the bank could implement non-disclosure agreements, ensuring that any adults coming into contact with sensitive data are legally bound not to share it. In this instance, however, the bank had failed to acquire the necessary agreements. This serves as another example of how confidential information can inadvertently be exposed.

These stories underscore the critical role humans play in preventing information leakage. By fostering awareness, implementing robust policies, and ensuring compliance across all levels, organizations can mitigate the risks associated with human error and fortify their defenses against potential breaches.

Information Leakage Protection Education

To mitigate the risks associated with dumpster diving and information leakage, consider implementing the following strategies:

- Secure Document Disposal: Establish clear policies and procedures for the disposal of sensitive information. Implement secure document disposal methods, such as shredding or incineration, to render information unreadable and unrecoverable. Ensure that employees are aware of and adhere to these procedures consistently.

- Data Classification and Retention Policies: Classify data based on its sensitivity and establish retention policies that outline how long data should be retained and when it should be securely destroyed. Regularly review and update these policies to align with legal and compliance requirements.

- Digital Transformation and Reduction of Physical Documents: Embrace digital transformation initiatives to minimize the volume of physical documents containing sensitive information. Digitize documents whenever possible, storing them securely and implementing strong access controls.

- Employee Training and Awareness: Educate employees about the risks of information leakage through dumpster diving and the importance of proper disposal practices. Provide training on how to identify sensitive information, handle it securely, and dispose of it using approved methods.

- Secure Trash Receptacles: Implement secure trash receptacles that deter unauthorized access. These can include locked bins or containers that prevent outsiders from retrieving discarded materials.

- Regular Auditing and Monitoring: Conduct regular audits and monitoring of information disposal practices. This includes reviewing disposal procedures, examining waste management contracts, and verifying compliance with established policies.

- Physical Security Measures: Implement physical security measures around waste disposal areas, such as restricted access, surveillance cameras, and adequate lighting. This helps deter unauthorized individuals from accessing dumpsters or trash bins.

- Information Security Culture: Foster a culture of information security within your organization. Encourage employees to take responsibility for protecting sensitive information and report any incidents or suspicious activities promptly.

By implementing these strategies and raising awareness about the risks associated with dumpster diving, organizations can significantly reduce the potential for information leakage and mitigate the risks of dumpster diving. However, it's important to remember that information security is an ongoing effort that requires vigilance and a collective commitment from everyone within the organization.

- Vendor and Partner Management: Extend your information security practices beyond the boundaries of your organization. Ensure that your vendors and partners also adhere to secure disposal practices for any sensitive information they handle on your behalf. Incorporate clauses related to information security in contracts and agreements to ensure compliance.

- Regular Risk Assessments: Conduct regular risk assessments that include an evaluation of information disposal practices. Identify any vulnerabilities or gaps in your current processes and implement necessary improvements. Keep up-to-date with emerging threats and evolving techniques used by dumpster divers to adjust your security measures accordingly.

- Employee Engagement and Reporting: Encourage employees to be actively engaged in maintaining information security. Establish a reporting mechanism for suspicious activities or observed lapses in disposal practices. Foster a culture where employees feel empowered to raise concerns without fear of retribution.

- Regulatory Compliance: Stay informed about relevant laws and regulations pertaining to information disposal in your jurisdiction. Ensure that your organization's practices align with these requirements to avoid legal consequences and penalties.

Remember, the risks associated with dumpster diving go beyond physical documents. In today's digital age, information leakage can also occur through discarded electronic devices or storage media. Therefore, it's crucial to implement proper disposal methods for electronic devices, including secure data wiping or physical destruction.

In the next chapter, we will explore another critical aspect of cybersecurity: the importance of user awareness and training. We will delve into effective techniques for educating employees about cyber threats, promoting a security-conscious mindset, and empowering them to be the first line of defense against attacks.

Stay alert, stay secure, and together, let's safeguard our valuable information.

Chapter 8: Education Techniques and Training Guides: Strengthening Cybersecurity Knowledge

In today's rapidly evolving cyber threat landscape, it is imperative to equip individuals and organizations with the knowledge and skills necessary to protect against attacks. Chapter 8 focuses on education techniques and training guides that aim to strengthen cybersecurity knowledge. By implementing effective training programs and providing practical guidance, readers can empower their teams to become the first line of defense against cyber threats.

In the ever-evolving landscape of cybersecurity, we often encounter a familiar scene: It's the dawn of a new year, and the HR/Compliance department reaches out to the entire workforce via email, signaling the start of the annual cybersecurity compliance training. Employees diligently log in, dedicating 30 minutes to an hour of their time to engage in a comprehensive course that delves into essential cybersecurity topics such as the insidious nature of phishing attacks and the art of safe browsing. With a final assessment, employees complete their cybersecurity awareness compliance training for the year. Throughout the months that follow, they may occasionally encounter a simulated phishing email or two, serving as gentle reminders of the ongoing vigilance required. However, these fleeting encounters barely scratch the surface of the intricate cybersecurity landscape that awaits. To set up a more meaningful and lasting impression on employee awareness, consider the following steps:

- Assessing Training Needs and Objectives:
  Before designing a training program, it is essential to assess the specific needs and objectives of the organization. Identify the knowledge gaps and areas of vulnerability to tailor the training to address those specific needs effectively.

- Designing a Comprehensive Training Plan:
  Develop a comprehensive training plan that outlines the curriculum, timeline, and resources needed. Consider the target audiences within the organization, such as employees from different departments, executives, and IT staff, and create training modules that cater to their specific roles and responsibilities.

- Choosing Training Methods and Approaches:
  Diversify training methods and approaches to accommodate different learning styles. Incorporate a mix of classroom training, online modules, gamification, and role-playing exercises to engage learners and ensure effective

knowledge transfer. Use real-life examples and case studies to make the training relatable and practical.

- Creating Engaging Training Content:
  Craft training materials that are informative, engaging, and easy to understand. Present complex cybersecurity concepts in a clear and concise manner. Utilize multimedia elements such as videos, infographics, and interactive content to enhance learning and retention.

- Reinforcing Training and Sustaining Knowledge:
  Training should not be a one-time event but an ongoing process. Implement regular assessments and knowledge checks to reinforce learning. Provide additional resources and encourage self-study to foster continuous improvement. Foster a culture of knowledge sharing and participation in cybersecurity communities and forums.

- Tailoring Training for Different Threat Vectors:

     Summarize the key points from previous chapters that discussed various threat vectors, including phishing, ransomware, SMishing/Vishing, DDoS, in-person social engineering, penetration testing, and dumpster diving. Highlight specific training considerations and best practices for each threat vector, emphasizing their interconnectedness.

- Evaluating Training Effectiveness:

     Establish metrics and indicators to evaluate the effectiveness of training programs. Measure knowledge retention, employee engagement, and the ability to apply learned skills. Gather feedback from participants and make necessary adjustments and improvements based on the evaluation results.

- Creating a Cybersecurity Awareness Culture:

     Develop a cybersecurity awareness culture within the organization. Engage leadership support and encourage a top-down approach to cybersecurity.

Foster employee involvement and accountability by encouraging reporting of suspicious activities. Integrate cybersecurity into organizational policies and procedures.

Chapter 8 highlights the significance of education techniques and training guides in strengthening cybersecurity knowledge. By implementing effective training programs, organizations can empower their teams to become proactive defenders against cyber threats. With a culture of continuous learning and a well-trained workforce, organizations can build robust cybersecurity defenses to safeguard their assets.

By following the guidance provided in this chapter, readers will be equipped to design and implement comprehensive training programs that enhance cybersecurity awareness and knowledge within their organizations. This will foster a security-conscious culture where individuals can effectively identify and respond to potential cyber threats, reducing the organization's overall risk.

Enhancements: For more information, including templates to assist in developing a comprehensive Cybersecurity Awareness program please visit https://www.thepeopleproblemproject.com

Conclusion

In the realm of cybersecurity, it is often said that humans are the weakest link. No matter how advanced our technological defenses may be, the actions and decisions of individuals within an organization can make or break its security. In this book, "The People Problem," we have explored the critical role that proper training plays in fortifying cybersecurity measures and mitigating the risks posed by various attack vectors.

Through personal stories, statistical insights, and practical strategies, we have journeyed through the world of cybersecurity, shining a light on the human factor and its impact on the security posture of organizations. We delved into attack vectors such as phishing, ransomware, smishing/vishing, DDoS attacks, in-person social engineering, and the often-overlooked risk of dumpster diving.

One recurring theme throughout this book has been the need for comprehensive and consistent training. It is not enough to invest heavily in cutting-edge cyber tools and technologies. To truly strengthen cybersecurity, organizations must prioritize the education and awareness of their employees. By instilling a security-conscious mindset, equipping individuals with the knowledge to recognize and respond to threats, and fostering a culture of vigilance, organizations can significantly reduce their vulnerability to attacks.

We have explored the power of personal stories, weaving together real-life experiences to highlight the impact that cybersecurity breaches can have on individuals and organizations. These stories serve as reminders that the consequences of inadequate training and awareness are far-reaching, affecting not only the bottom line but also the trust and confidence of stakeholders.

To complement these personal stories, I have presented compelling statistics and data that underscore the urgency of addressing the people problem in cybersecurity. From the staggering financial losses resulting from data breaches to the prevalence of social engineering tactics, the numbers speak volumes about the risks organizations face.

But this book is not meant to instill fear or despair. Instead, it aims to empower. I have provided actionable strategies and techniques to strengthen cybersecurity through proper training. From establishing a robust security awareness program to conducting regular penetration testing and promoting a culture of information security, the tools are at our disposal. By implementing these measures, organizations can enhance their resilience, minimize the likelihood of successful attacks, and respond effectively in the face of evolving threats.

However, the journey does not end with the conclusion of this book. Cybersecurity is an ever-changing landscape, requiring continuous adaptation and improvement. The strategies and techniques presented here are a starting point, but they must be accompanied by ongoing efforts to stay informed, educate employees, and remain vigilant.

Remember, cybersecurity is a collective responsibility. It is not solely the domain of IT departments or security professionals. It is the responsibility of every individual within an organization. By working together, supporting one another, and fostering a culture of security, we can build stronger defenses and protect the invaluable assets under our care.

"The People Problem" highlights the importance of investing in the human element of cybersecurity. It serves as a call to action for organizations to prioritize proper training, empower their employees, and recognize that people are not just the weakest link—they can also be the first line of defense against cyber threats.

Let us embark on this journey together, armed with knowledge, resilience, and a commitment to safeguarding our digital world.

Acknowledgements

If you have ever tried to undertake the daunting task of writing a book, you understand that even though there may be only one author listed under the title there are several people responsible for getting this book into production. I would like to take a few paragraphs to thank the many people who have helped make this work possible.

I honestly thank God for providing the different (& often broken) roads that I have traveled down to obtain the knowledge and experience necessary to include in this book. With his guidance and providence all things are possible.

Additionally, I want to thank my wife, Amanda and three kids, Addison, Kyndall and Beckham. During the time I worked as a consultant, I worked away from home, sometimes many weeks at a time. They held down the fort and understood why I had to be gone. Without their patience and understanding, I would have never been able to experience the personal stories shared in this book.

I would be remiss without thanking some of the leaders in my cybersecurity career who have always encouraged me to do my best. Thank you, Russ Horn, for being a Godly example of a leader by showing me how to be a compassionate auditor and building a good rapport with those companies I worked with. Thank you, Rusty Jordan, for being a kind of leader who leads by implementing ideas because they make sense, not necessarily because they are the flashiest or most expensive.

Finally, thank you to my supporters like Josh and Natalie. Your encouragement and support in helping me get this book across the finish line will not soon be forgotten.

It is my sincere hope that those who read this book will be inspired to properly train their staff and create a cyber aware culture as our people are truly the last line of defense in our cyberwarfare.

www.ingramcontent.com/pod-product-compliance
Lightning Source LLC
Chambersburg PA
CBHW060352130626
46553CB00003B/1201